Materialien für den bilingualen Unterricht
Biologie

Your Respiration

How and Why You Breathe

Cornelsen

Materialien für den bilingualen Unterricht: Biologie
Your Respiration – How and Why You Breathe

Erarbeitet von
Horst-Dieter Mathews

Abgeleitet aus dem Kapitel „Respiration“ im bilingualen Lehrwerk
Discover Biology – Volume 1 (ISBN 978-3-464-31844-7)

Beratende Mitwirkung: Simon Olmesdahl
Redaktion: Marc Proulx, Mareike Seehaus (Assistenz)
Umschlaggestaltung: Klein & Halm Grafikdesign, Berlin
Layoutkonzept: werkstatt für gebrauchsgrafik, Berlin
Technische Umsetzung: Rainer Bachmaier

www.cornelsen.de

1. Auflage, 1. Druck 2009

Alle Drucke dieser Auflage sind inhaltlich unverändert und können im Unterricht nebeneinander verwendet werden.

Druck: H. Heenemann, Berlin

ISBN 978-3-06-032327-2

Inhalt gedruckt auf säurefreiem Papier aus nachhaltiger Forstwirtschaft.

Your Respiration – How and Why You Breathe is a bilingual module for use in your biology class. With this module I think you will discover that learning about biology in English does not have to be difficult. In fact, you may even find it more enjoyable than learning in your own language.

Helpful tools in this module

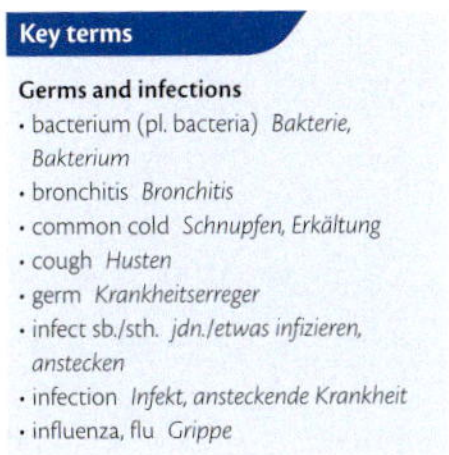
Key terms
Germs and infections
- bacterium (pl. bacteria) *Bakterie, Bakterium*
- bronchitis *Bronchitis*
- common cold *Schnupfen, Erkältung*
- cough *Husten*
- germ *Krankheitserreger*
- infect sb./sth. *jdn./etwas infizieren, anstecken*
- infection *Infekt, ansteckende Krankheit*
- influenza, flu *Grippe*

Key terms[1]

At the beginning of each new section you will find a box with *Key terms*. These are important words used in biology which appear for the first time in the module. Always look at the key terms before reading the text, then watch out for them while you are reading. You can find the terms easily in the text because they are in **bold print**.

New vocabulary

Words and phrases which you might not know yet are numbered[2] in the text and explained in German on the same page.

Pronunciation help

If you are not sure how to pronounce a key term or a new word, you can find out by looking up the word in the appendix at the back of the module (see also 'Appendix' below).

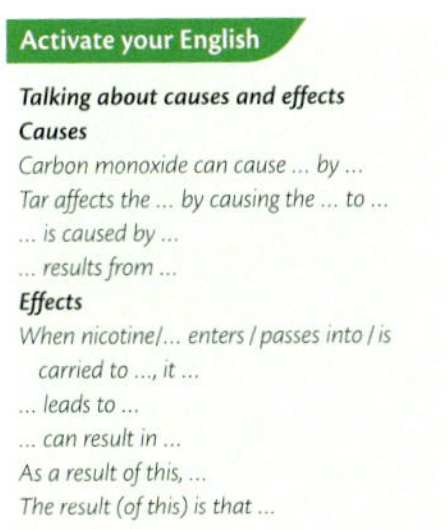
Activate your English
Talking about causes and effects
Causes
Carbon monoxide can cause ... by ...
Tar affects the ... by causing the ... to ...
... is caused by ...
... results from ...
Effects
When nicotine/... enters / passes into / is carried to ..., it ...
... leads to ...
... can result in ...
As a result of this, ...
The result (of this) is that ...

Tasks and *Activate your English*

The tasks at the end of most sections help you to understand the information in the texts and pictures and make you think further about the material. They also give you practice in discussing topics and ideas with other people. To help you say what you want to say in English, you will find words and phrases under the heading *Activate your English* in a box next to a task.

Saying it in German

You will find at least one task or activity with the heading 'Say it in German', which asks you to speak or write in German about something you have just learned. These exercises test your understanding of the material and give you practice using the equivalent terminology in German.

Biology skills[3]

The coloured pages introduce a biology skill that is relevant to the work in the module. These skills pages offer you strategies and methods to help you learn better and more easily. Tasks and activities give you practice with the new skill.

Revision

A revision page at the end of the module allows you to review and practise the things you have learned. You can also find exercises to improve your English.

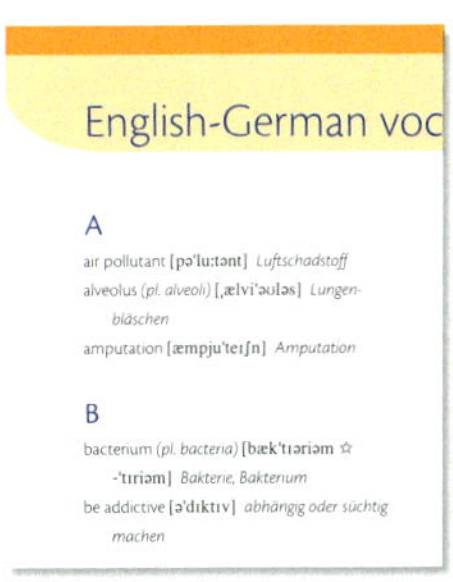
English-German voc
A
air pollutant [pə'lu:tənt] *Luftschadstoff*
alveolus *(pl. alveoli)* [ˌælvi'əʊləs] *Lungenbläschen*
amputation [æmpju'teɪʃn] *Amputation*
B
bacterium *(pl. bacteria)* [bæk'tɪəriəm ☆ -'tɪriəm] *Bakterie, Bakterium*
be addictive [ə'dɪktɪv] *abhängig oder süchtig machen*

Appendix[4]

In this last section all the key terms and annotations are listed with their German equivalents in an English-German vocabulary list. If you have trouble pronouncing a word in English, here you can look up the phonetic spelling of the word. A German-English vocabulary list provides help when you know what you want to say in German but cannot find the words in English.

I hope you enjoy working with this module!

Horst-Dieter Mathews

1 key term *Schlüsselwort*
2 number sth. *etwas nummerieren*
3 skill *Fertigkeit*
4 appendix *Anhang*

Your Respiration

How and Why You Breathe

Objectives

- To **name** the parts of your respiratory system and **describe** their function
- To **explain** the role of the diaphragm and the intercostal muscles in breathing
- To **explain** how the gases in the air are exchanged in your lungs
- To **find out** the effect of exercise on your breathing rate
- To **analyse** the dangers of smoking to a person's health

Since the moment you came into this world, you have been breathing in and out, day and night, even when you sleep. This process is called **respiration**: your body takes in **oxygen** from the atmosphere and releases **carbon dioxide**. But why does your chest and your belly move when you breathe? Why do we have to breathe at all? And where does the oxygen go that you take in? This module will help answer these questions.

1 The organs used for breathing make up your respiratory system

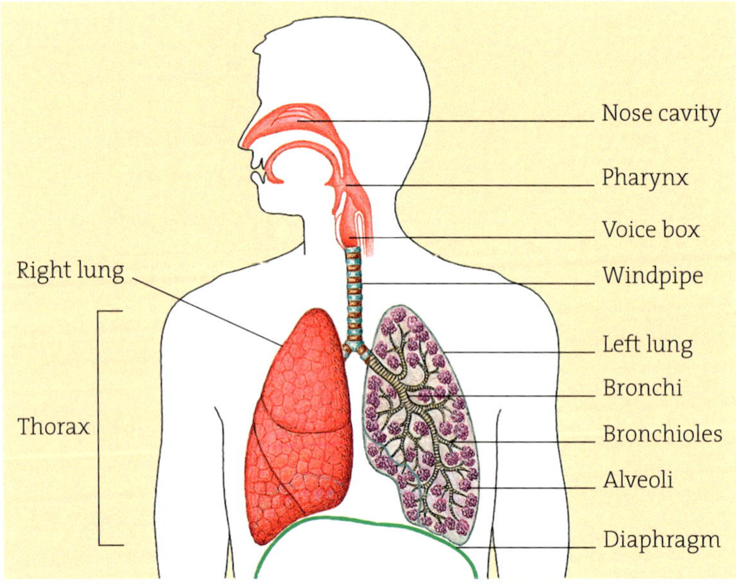

Figure 1 The respiratory system

Key terms

Anatomy of the lungs
- alveolus (pl. alveoli) *Lungenbläschen*
- bronchus (pl. bronchi) *Bronchie*
- bronchiole *Bronchiole*
- capillary *Kapillar*
- cilium (pl. cilia) *Wimperhärchen*
- diaphragm *Zwerchfell*
- left and right lungs *Lungenflügel*
- mouth cavity *Mundhöhle*
- nasal cavity *Nasenhöhle*
- pharynx *Rachen*
- pleural membrane *Lungen- und Rippenfell*
- voice box *Kehlkopf*
- wind-pipe (trachea) *Luftröhre*

Breathing (1)
- carbon dioxide *Kohlenstoffdioxid*
- oxygen *Sauerstoff*
- respiration *Atmung*
- respiratory system *Atmungssystem*

Let us first look at how air gets into your lungs. When you breathe in, the air enters your body through your nose or your mouth. In your **nasal cavity** the air is warmed, moistened[1] and cleaned. The inside of your nose is covered in mucus[2]. Here dust particles and germs[3] are caught and millions of tiny hairs (**cilia**) brush them towards your throat. By sneezing and coughing, your body forces them back out.

The air then passes into your **pharynx** and through the **voice box**, which helps us produce sounds for speaking and singing. It flows down your **windpipe (trachea)**, which is 11 to 14 cm long, and divides[4] into two tubes called **bronchi**. Rings of cartilage[5] protect the windpipe and hold it open all the time. The bronchi carry the air into both lungs. The lungs are spongy[6], shaped like sacks and lie inside the rib cage, which protects them (see Figure 2).

1 moisten sth. *etwas anfeuchten*
2 mucus *Schleim*
3 germ *Bazille, Bakterie*
4 divide into sth. *sich in etwas aufteilen*
5 cartilage *Knorpel*
6 spongy *schwammig, weich*

The bronchi then divide into smaller and smaller tubes, or **bronchioles**, like the branches[1] and twigs[2] of a tree. At the ends of the bronchioles are clusters of tiny balloons called **alveoli**. Each individual alveolus is smaller than a grain of salt. Around each one is a close net of blood **capillaries** (see Figures 3 and 4). The purpose and function of the alveoli and capillaries will be explained in the next section.
Between the lungs and ribs are two layers[3] of skin called **pleural membranes**. They allow the lungs to move smoothly along the ribs. A thin sheet of muscle, the **diaphragm**, separates your thorax (the part of the upper body which includes the ribs, the heart and lungs) from the organs of your abdomen.

1 Explain why it is better to breathe through your nose than through your mouth.
2 What do you think would happen if there were no rings of cartilage round the windpipe?
3 Describe the function of the pleural membranes.

2 The lungs cannot work without muscles

Key terms

Breathing (2)
- contract *sich zusammenziehen*
- intercostal muscles *Zwischenrippenmuskeln*
- relax *sich entspannen*

Your lungs are soft and cannot move by themselves. So how does breathing work?
When you watch somebody run or play sports, you can see that his or her chest moves up and down. This is because there are muscles between the ribs called **intercostal muscles**, which **contract** to lift the ribcage. This makes more space inside the ribcage so that the lungs can expand and fill with air. When the intercostal muscles **relax**, the ribcage sinks, the space inside becomes smaller, and the air is pushed out of the lungs.
Another muscle plays an important role during breathing: the **diaphragm**. It is normally dome-shaped[4], but when you breathe in, it contracts and flattens, making the room inside your chest bigger and sucking air into your lungs. When the diaphragm relaxes, it becomes dome-shaped again and the space for the lungs becomes smaller. This forces the air out of the lungs.

1 branch *Ast, Zweig*
2 twig *dünner Zweig*
3 layer *Schicht*
4 dome-shaped *gewölbt*

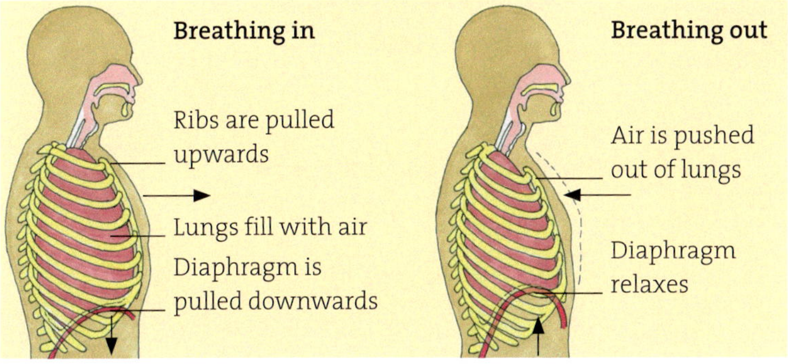

Figure 2 The movement of the ribcage and diaphragm

When you look at your classmate sitting next to you, you can hardly see any breathing movement in his or her chest. So how do we breathe when we are at rest?
In this case, the contracting and relaxing of the diaphragm muscle brings enough air into your lungs to give the organs the oxygen they need.
You will discover more about the working of your diaphragm when you do the bell jar experiment (see Improve your skills, p. 10).

1 Stand up straight and place your left hand on your lowest ribs and your right hand flat on your abdomen. Then breathe in and out normally. Describe and explain how your ribcage moves.
2 How do you breathe when you are sleeping?
3 **Say it in German:** Explain to a friend how his/her lungs work at rest and during sport.

3 Gas exchange takes place in the alveoli and in the cells

Alveoli and capillaries both have very thin walls so that gases can pass through them easily. The air you breathe in has a lot of oxygen in it, which passes through the capillary walls and enters the bloodstream[1]. At the same time, the blood coming from the organs carries carbon dioxide. So while oxygen is being taken into the body, carbon dioxide is being moved in the opposite direction: it passes from the bloodstream into the alveoli and is then breathed out. This process is called **external respiration**.

The oxygen coming in is taken up by the red blood cells in the capillaries, which carry it to the different cells of your body. Those cells then use the oxygen to get **energy** from food substances you have eaten, which the blood has also transported there. The oxygen is needed to burn these substances in a process which produces carbon dioxide. Since this process takes place inside your cells, it is called **internal respiration**, also known as cellular respiration.

Key terms

Exchange of gases
- energy *Energie*
- external respiration *äußere Atmung*
- internal respiration *innere Atmung*

Did you know?

There are about 300 million alveoli in your lungs. If they were placed side by side, their combined surface would form an area of about 70 square metres – nearly the size of half a tennis court.

1 bloodstream *Blutbahn, Blutkreislauf*
2 close-up *Nahaufnahme*
3 nitrogen *Stickstoff*

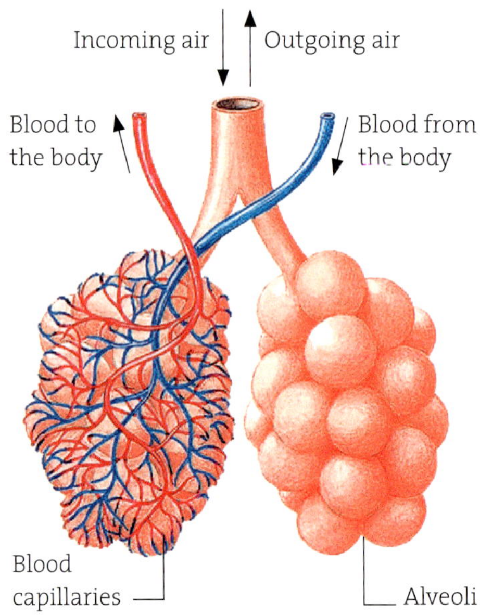

Figure 3 Alveoli with capillaries

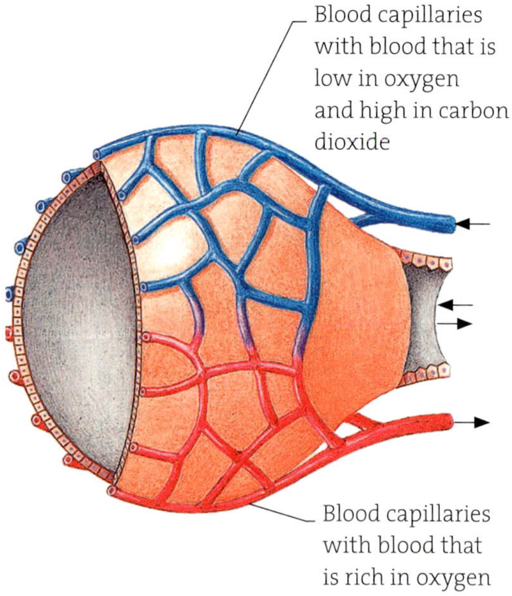

Figure 4 Close-up[2] of alveolus

1 Describe where gases are exchanged in the body and in which direction this exchange takes place.
2 What is the advantage of having such a large number of small alveoli?
3 Look at the table on the right and answer these questions:
   a How does the amount of oxygen and carbon dioxide change in your lungs?
   b Why do we breathe out more carbon dioxide than we breathe in while the amount of nitrogen stays the same?

Gas	% in air you breathe in	% in air you breathe out
Oxygen	21	16–17
Carbon dioxide	0.03	3.5
Nitrogen[3] and other gases	78	78

4 Smoking is bad for your health

Health organizations have found that children have their first contact with cigarettes between the ages of 8 and 14. Children and teenagers start smoking for different reasons. Some start because their friends smoke, while others want to impress people or be accepted in a group. The earlier young people start smoking, the greater are the dangers to their health.

In Germany more than 112,000 people die each year from diseases connected with smoking. This is about the number of deaths that would occur if a large plane crashed every day and killed all its passengers. According to the World Health Organization, about half of all regular cigarette smokers are killed by their habit[1].

Figure 5 He wants to belong to the group

What makes tobacco smoke so dangerous?

In the European Union every packet of cigarettes and tobacco must display a health warning in bold letters and information on how much tar, nicotine and carbon monoxide are contained in one cigarette. This is important information because these three substances are the reason why smoking is so dangerous.

Nicotine is a poisonous[2] **drug**, which passes quickly from the lungs into the blood and is carried to all the organs in the body. It has two particularly negative effects on the body: First, nicotine is **addictive**. When the body gets used to[3] nicotine, it does not want to be without it. This is why many people find it hard to give up smoking. Second, the drug causes your blood vessels[4] to contract, which makes it difficult for your blood to circulate properly. This can lead to **heart attacks** and **strokes**. The risk of having a heart attack is two or three times higher in smokers than in non-smokers. Also, most leg **amputations** in Germany are carried out on smokers. The tissues in the arms and legs can die when they do not get enough oxygen-rich blood.

Carbon monoxide, which is one of over 4000 **chemical compounds**[5] in tobacco smoke, is also absorbed into the blood. This gas not only reduces the amount of oxygen that the blood can transport, but also affects body cells so that they cannot take in as much oxygen as they normally do. This is why smokers are often short of breath[6] when they are active during work or sports. Carbon monoxide is also produced by car engines, so it not only affects smokers.

Key terms

Smoking and its effects

- amputation *Amputation*
- be addictive *abhängig oder süchtig machen*
- bronchitis *Bronchitis*
- cancer *Krebs, Krebserkrankung*
- carbon monoxide *Kohlenstoffmonoxid*
- drug *Droge, Arzneimittel*
- emphysema Emphysem
- headache *Kopfschmerz*
- heart attack *Herzanfall*
- inflammation *Entzündung*
- lung cancer *Lungenkrebs*
- nicotine *Nikotin*
- passive smoking *Passivrauchen*
- stroke *Schlaganfall*
- tar *Teer*
- tumour *Geschwulst, Tumor*

1 habit *Sucht*
2 poisonous *giftig*
3 get used to sth. *sich an etwas gewöhnen*
4 blood vessel *Blutgefäß*
5 chemical compound *chemische Verbindung*
6 breath *Atem*

Tar is the third dangerous substance and the reason why the lungs of heavy smokers are coloured black on the inside. When tobacco smoke cools in the mouth, pharynx and lungs, the tar in the smoke settles[1] on the inner surface of these organs. Someone who smokes 20 cigarettes a day breathes a full cup of tar (210 g) into his or her lungs in one year.

Normally your lungs are kept clean by mucus, which covers the inner lining[2] of the lungs. The mucus catches and holds dust particles and germs. The cilia (see p. 4) then transport the mucus out of the lungs. Tar makes the cilia stick together so that they stop working properly. The mucus then stays in the lungs and becomes an ideal breeding ground[3] for bacteria and viruses. It can also lead to a disease called **emphysema** when the affected person has to cough repeatedly to get rid of[4] the mucus. This damages the delicate alveoli. Holes then develop in the lung tissue (see Figure 6) and it becomes more difficult for the person to breathe.

Figure 6 Lung tissue showing signs of emphysema

Tar and smoke also irritate the lining of the respiratory organs. When the bronchi become swollen and painful this **inflammation** of the bronchi is called **bronchitis**. Finally, tar contains over 400 different chemicals, and more than 30 of them can cause **cancer**.

Lung cancer is the most serious disease which can result from long-time smoking. Cancer begins when cells of the inner lining of the bronchi start to divide in an uncontrolled way. Later **tumours** develop in the lungs. Once lung cancer starts, it may spread to other parts of the body via the blood and lymph system. Lung cancer is difficult to detect at an early stage, when it can still be treated. If smokers go to see their doctor when they feel a pain in their chest, it may be too late and the chances of surviving are relatively small. In Germany, Britain and the USA, lung cancer is one of the major causes of death. In Germany, 54,000 people die of lung cancer every year. Although some non-smokers may get lung cancer, 80 per cent of all people who die of this disease are smokers.

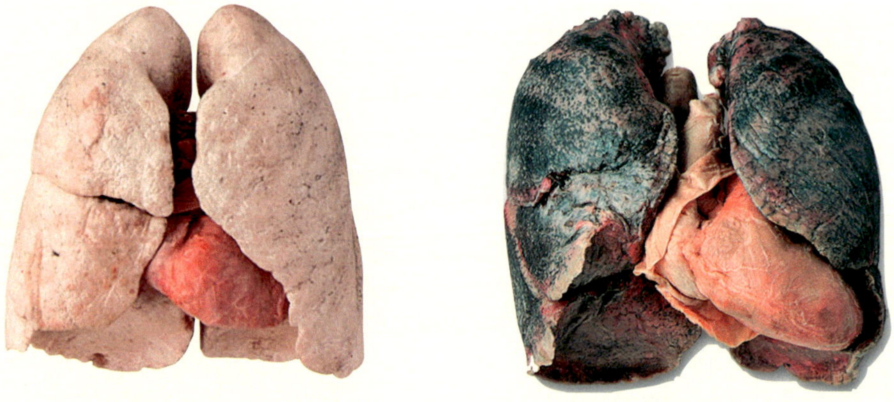

Figure 7 A healthy lung (left) and the lung of a heavy smoker (right)

Smoking is especially dangerous during pregnancy[5]. If a woman smokes during her pregnancy, the chances are high that her new-born baby will be smaller than normal because the develop-

1 settle *sich niederlassen*
2 inner lining *Schleimhaut*
3 breeding ground *Brutstätte*
4 get rid of sth. *etwas loswerden*
5 pregnancy *Schwangerschaft*

ing foetus[1] will get less oxygen and food. Brain cells are especially sensitive to carbon monoxide, so the risk of damaging the foetus's brain is greater.

Even people who do not smoke at all can be affected by tobacco smoke if they are around smokers too often. **Passive smoking** not only irritates your eyes and causes sore throat[2] and **headaches**, it can lead to the same diseases that smokers get. Today more and more countries are introducing laws which forbid smoking in public buildings, hospitals, offices and restaurants, in order to protect the health of non-smokers.

1 foetus *Fötus*
2 sore throat *Halsschmerzen, Halsentzündung*

State	Bars	Restaurants	Private workplaces other than bars or restaurants	Government buildings
Alabama			▲	▲
Alaska		▲	▲	▲
Arizona	●	●	●	●
California	●	●	●	●
Florida		●	●	●
Nevada		●	●	●
New York	●	●	●	●
Texas				
Wyoming				▲
England	●	●	●	●

● Smoking is prohibited ▲ Smoking is restricted

Table 2 **Anti-smoking laws in some US states and in England** *(Source: http://ash.org/smokingbans.html, 25 April 2008)*

1 Name the three most dangerous substances in tobacco smoke. In small groups, put the information on these substances from pp. 7–8 into cause-and-effect diagrams as shown below (one diagram for each substance). Then take turns describing the effects of the substances to each other. See **Activate your English** below for language help.

Cause	*Effect(s)*
Nicotine enters the blood ...	...

2 Explain why there is even more reason for pregnant women not to smoke.

3 In Germany the average age when people start smoking is 13 ½. Why do you think teenagers take up smoking? Make a list of reasons why it is better not to start.

4 **a** Look at Table 2 and compare and contrast the laws in two US states or in one of the states and England. The following words and phrases will help you: *... is more/less strict than ... – ... is not as strict as ... – ... is (not) prohibited/restricted in ... – ... are (not) allowed to smoke in ... – Although, ... – However, ...*

b Research: Find out what the current anti-smoking laws are where you live and compare them with the laws in England.

Activate your English

Talking about causes and effects

Causes

Carbon monoxide can cause ... by ...
Tar affects the ... by causing the ... to ...
... is caused by ...
... results from ...

Effects

When nicotine / ... enters / passes into / is carried to ..., it ...
... leads to ...
... can result in ...
As a result of this, ...
The result (of this) is that ...

Working with models

Models are helpful in biology for understanding complicated processes. They are effective because they only show the most important features of what is being described or simulated. By leaving less important aspects out, models make things easier to understand. This also means, however, that they do not fully represent reality. We need to keep this in mind when we compare models with 'the real thing'. The following examples demonstrate this.

1 Using a bell-jar[1] model

Materials:

- a plastic bell jar (see Figure 8 below)
- a glass tube[2] branched in two at one end
- a rubber sheet[3]
- a rubber ring for fastening the rubber sheet
- a rubber bung[4] with one hole
- two balloons

1 bell jar *glockenförmiges Gefäß*
2 glass tube *Glasrohr*
3 rubber sheet *Gummihaut*
4 rubber bung *Gummistopfen*

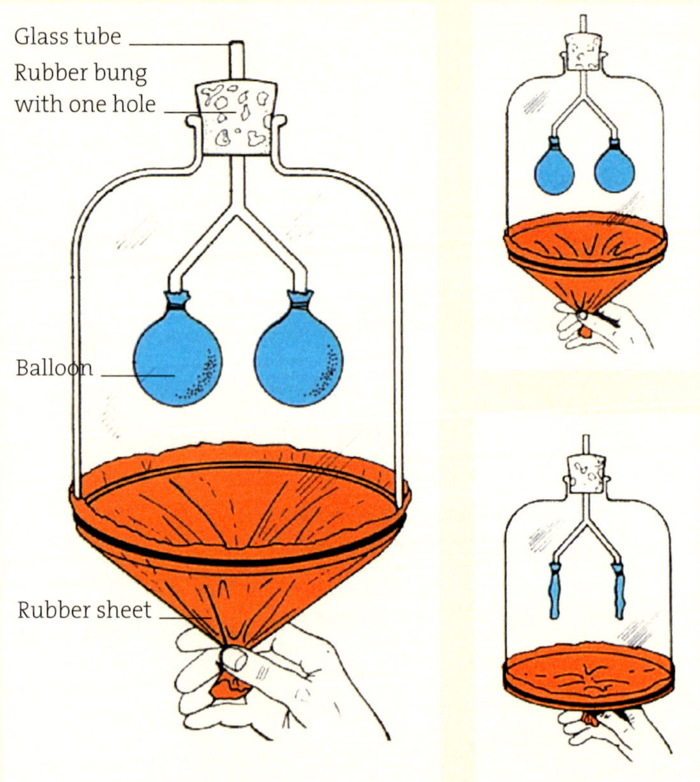

Figure 8 The bell-jar model

Procedure:

1. Pull the rubber sheet downwards (see Figure 8). What happens inside the bell jar?
2. Now push the rubber sheet upwards. Describe what happens.
3. Explain why the balloons inside the bell jar behave this way.
4. Now think about your lungs. Compare the model with your respiratory system by completing the following table.

Part of the model	*Respiratory organ*
Bell jar	...
...	...

Table 3 A comparison

1 Based on what you have observed from the bell-jar model, explain how we breathe in and out. Make sure you mention all the organs from the second column of the table above.

2 Explain why the model cannot show exactly how breathing works in our body.

3 How is our breathing during physical exercise different from the breathing demonstrated by the bell-jar model?

2 Making a smoking apparatus

The aim of this activity is to better understand what smoke does to your lungs.

Materials:
- Erlenmeyer flask[1]
- Glass tubes (two)
- Rubber bung with 2 holes
- Cigarette without a filter
- Rubber tubing[2]
- Suction bulb[3] or suction pump[4]
- Cotton wool[5]

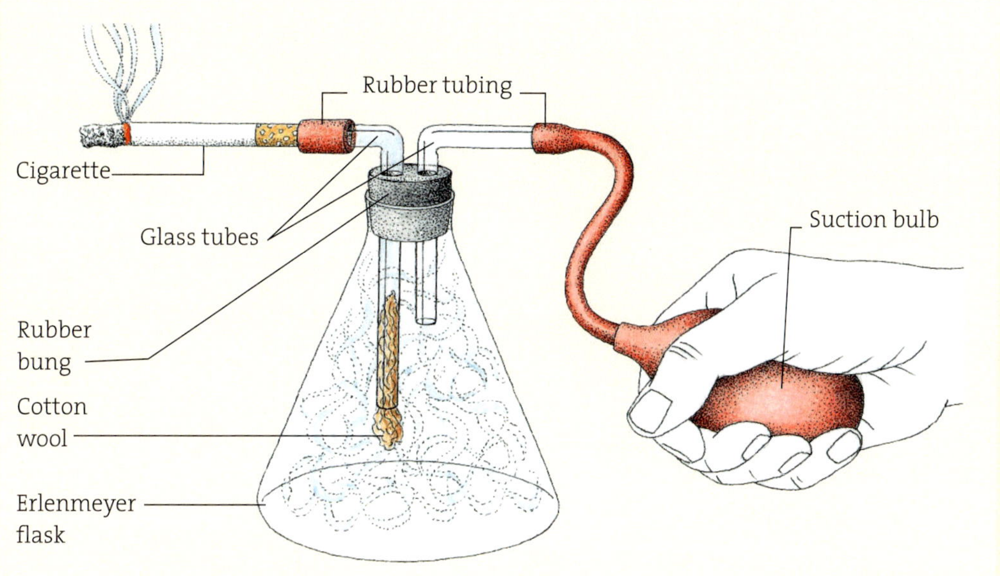

Figure 9 A smoking apparatus

Figure 10 A cigarette filter does not absorb all dangerous substances

Procedure:
1. Set up the apparatus as seen in the picture above. Use a short rubber tubing to connect the cigarette to the glass tube.
2. Light the cigarette. Then draw air[6] through it by squeezing the suction ball. Repeat the action for about one minute.
3. Describe what happens to the model as you work the suction ball.
4. Write down your observations.

1 Erlenmeyer flask *Erlenmeierkolben*
2 rubber tubing *Gummischlauch*
3 suction bulb *Saugball*
4 suction pump *Saugpumpe*
5 cotton wool *Watte*
6 draw air *Luft ziehen*

1 As in all models, certain parts of the apparatus stand for certain things in reality. Make a table and match each part with the corresponding organ in your body (see Table 3, p. 10). Note: the rubber tubing and bung are needed for the model and do not represent parts of the body.

2 Use your observations and what you have learned from the module so far to explain the effect that smoking has on your organs.

Activity 1

Investigating[1] how exercise affects the breathing rate[2]

Before you begin, read all the instructions. Then do the whole activity without stopping.

1. a Sit quietly and relaxed for about *two minutes*.
   b Now count for *one minute* how many times you breathe in.
   c Record[3] the number in your exercise book.
2. a Do some physical exercise[4], for example bending the knees or stepping up onto a chair and down again, for *two minutes* without stopping.
   b Now sit down quickly and count again for *one minute* how many times you breathe in.
   c Record the number in a table as shown in Table 4 below.
   d Relax for *one minute*, then record your breaths per minute again. Repeat this four times.

This diagram gives you an overview. Starting at the top, follow the wheel in a clockwise direction.

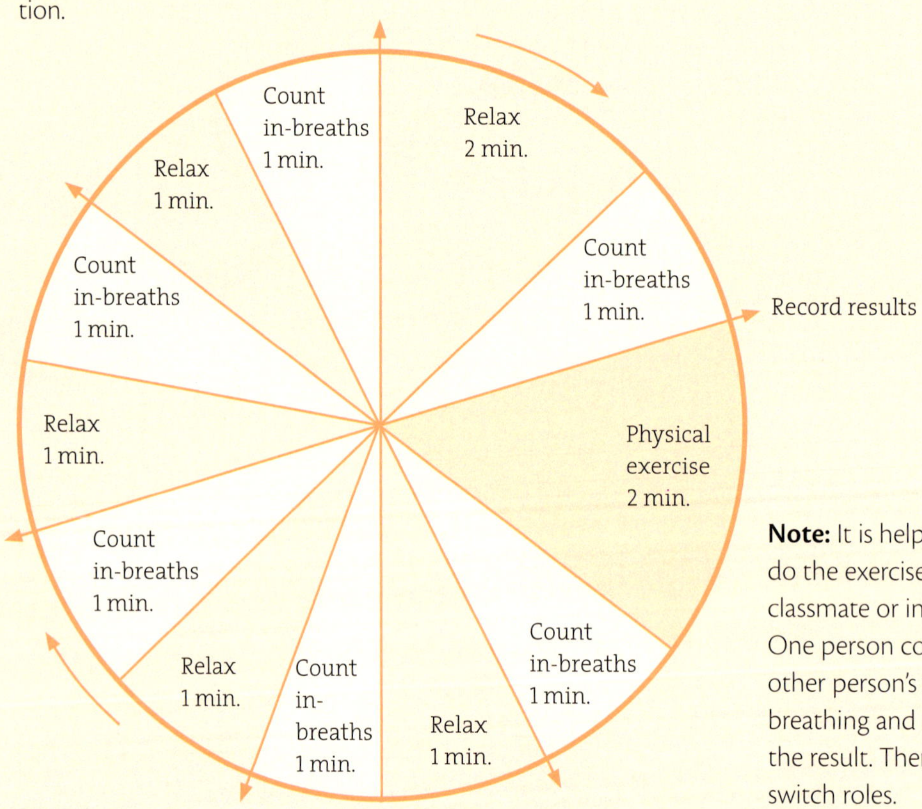

Note: It is helpful to do the exercise with a classmate or in a group. One person counts the other person's rate of breathing and records the result. Then you switch roles.

Figure 11 Experimenting with your breathing rate

Time passed after activity	*Number of breaths per minute*
1 min.	...
3 min.	...
5 min.	...

Table 4 Activity and breaths per minute

1 investigate sth. *etwas untersuchen*
2 breathing rate *Zahl der Atemzüge pro Zeiteinheit*
3 record sth. *etwas aufschreiben, notieren*
4 physical exercise *körperliche Aktivität(en)*

1 Compare your results with those of your classmates.
2 Give reasons for the quick rise in the number of breaths per minute during the exercise.
3 Why do you think it takes so long after exercising for your breathing rate to return to normal?

5 Breathing, swallowing and speaking

When you breathe in through your nose or mouth, the air passes first through the pharynx and then the windpipe or trachea. At the opening of the trachea is a small piece of tissue called the **epiglottis**, which directs air to the voice box. When you breathe in or out, the epiglottis is in an upright position so that air can pass through the trachea. When you swallow food, it automatically folds down and closes the entrance to the trachea so that the food goes into the gullet[1].
This is why it is not a good idea to talk or laugh while you are eating or drinking, as food or liquid can go 'down the wrong pipe'. When this happens, a reflex action in the voice box makes you cough until the food or liquid is forced out of the windpipe.

Key terms

Breathing and speaking
- epiglottis *Kehldeckel*
- vocal cords *Stimmbänder*

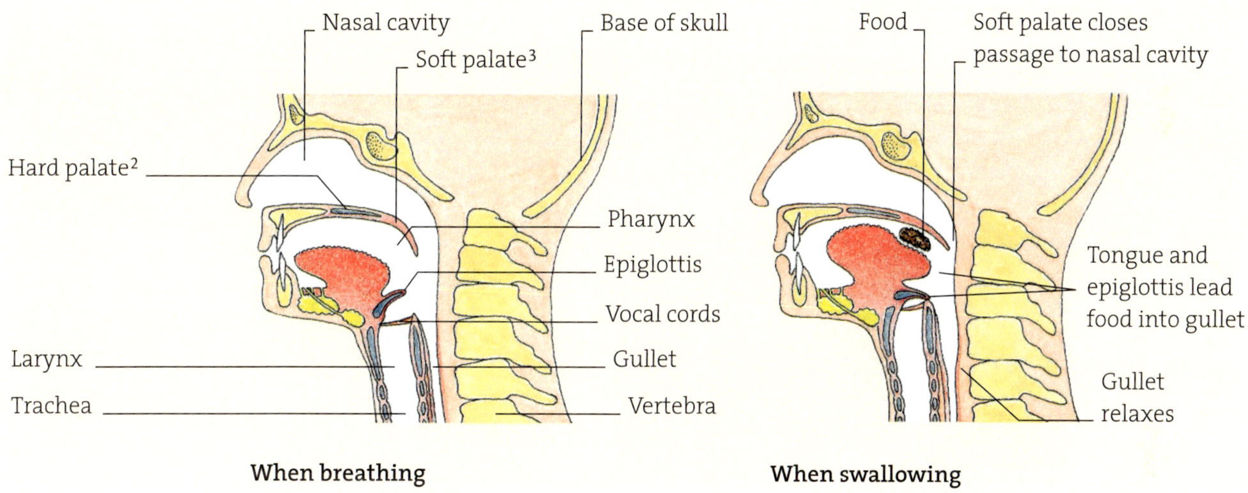

Figure 12 Position of the epiglottis

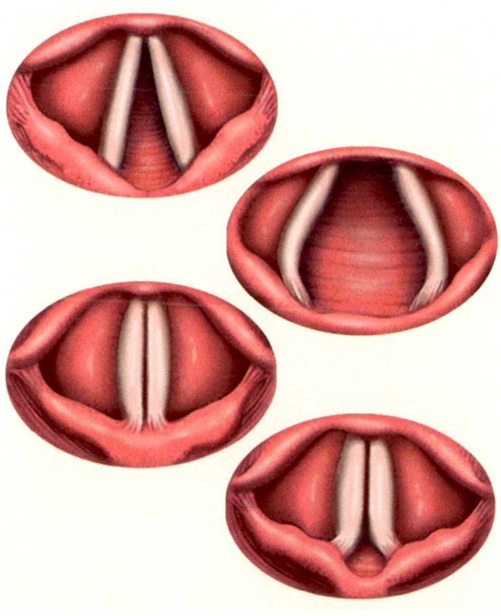

Figure 13 How vocal cords produce different sounds

Further facts

Your voice
At the front and back of the voice box are two flat folds of skin called the **vocal cords**. When air passes over them, they vibrate[4] and produce sounds. The pitch[5] and the quality of the sounds are varied when certain muscles change the tension[6] in the cords and the distance between them. As the vocal cords are slightly longer in males than in females, men normally have deeper voices than women. The human voice is produced by a complex coordination[7] of the lips, the tongue, the jaws, the breathing movements and the tension and position of the vocal cords.

1 gullet *Speiseröhre*
2 hard palate *harter Gaumen*
3 soft palate *weicher Gaumen, Gaumensegel*
4 vibrate *vibrieren*
5 pitch *Tonhöhe*
6 tension *Spannung*
7 coordination *Abstimmung, Koordination*

6 Diseases of the respiratory organs

Many diseases of the respiratory system are caused by air pollutants[1] from the burning of fossil fuels[2] in industrial plants[3], households, cars and other motor vehicles. These substances damage the breathing passages and make it easier for **germs** to **infect** the mucous membranes[4] in the nose, mouth, windpipe and lungs.

Common cold: This **infection** of the nasal passages[5] is caused by a **virus**. Typical symptoms are a running nose, light fever, headache, and loss of appetite and sense of smell. While no medication can stop a cold virus, some can relieve[6] the unpleasant symptoms. When you have a cold, you should keep your body warm and drink lots of liquids. After two or three days your immune system will normally win the fight against the germs and you will start feeling better.

Key terms

Germs and infections
- bacterium (pl. bacteria) *Bakterie, Bakterium*
- bronchitis *Bronchitis*
- common cold *Schnupfen, Erkältung*
- cough *Husten*
- germ *Krankheitserreger*
- infect sb./sth. *jdn./etwas infizieren, anstecken*
- infection *Infekt, ansteckende Krankheit*
- influenza, flu *Grippe*
- pneumonia *Lungenentzündung*
- sore throat *Halsschmerzen, Halsentzündung*
- tuberculosis *Tuberkulose*
- virus *Virus*

Figure 14 Inhalations help dissolve mucous

Cough: This condition[7] often arises in connection with a cold. When the lining of your throat and, in some cases, the upper part of your windpipe become infected, you get a **sore throat**. By coughing, the body throws out mucus and the harmful viruses and **bacteria** that gather there. If the cough does not go away after a few days, you should see your doctor.

Influenza: Better known as (the) **flu**, this viral infection has symptoms similar to a cold. Sore throat and aching muscles are also typical. The treatment[8] is basically the same: rest and drink lots of liquids, preferably water. A flu can be dangerous and even deadly for old or weak people. It can also sometimes lead to more serious diseases such as pneumonia.

Pneumonia and **tuberculosis:** Both are dangerous infectious diseases of the lungs which are caused by bacteria. They can be treated[9] with antibiotics.

Bronchitis: In a person with bronchitis the lining of the bronchi is inflamed[10] and secretes a thick and sticky mucus. The body tries to clear out the mucus by coughing. Inhalations[11], e.g. with dried chamomile flowers, wet packs[12] and the drinking of special teas helps dissolve the mucus and relieve the inflammation.

To protect yourself against infections: Do not work or play in rooms that are overheated, because the air dries out the mucous membranes of your nose and throat. Avoid breathing in tobacco smoke and the exhaust fumes of cars. Spend a lot of your free time outside, get exercise and eat lots of fresh fruit and vegetables every day.

1 air pollutant *Luftschadstoff*
2 fossil fuel *fossiler Brennstoff*
3 industrial plant *Industriebetrieb*
4 mucous membrane *Schleimhaut*
5 nasal passage *Nasengänge*
6 relieve *erleichtern*
7 condition *Befinden, Zustand*
8 treatment *Behandlung*
9 treat sb./sth. *jdn./etwas behandeln*
10 inflamed *entzündet*
11 inhalation *Inhalieren*
12 wet pack *feuchter Wickel*

1 Why do you think it is good to open classroom windows during breaks?
2 Compare the common cold with influenza.
3 When did you last have a cold? What were the symptoms and how did you get rid of it?

Activity 2 Measuring[1] the vital capacity[2] of your lungs

Materials:
- a large bottle (5–6 litres) with measurements (see Figure 15 below)
- a rubber stopper[3]
- a piece of rubber or plastic tubing
- a glass or plastic container filled with water
- a glass tube for a mouthpiece
- a glass tube for the other end of the rubber tubing
- a stand with a clamp[4] to hold the bottle

Procedure:
1. Fill the large bottle completely with water and close it with the rubber stopper. Place the bottleneck under water in the container, clamp[5] the bottle to the stand and loosen the rubber stopper. Then stick one end of the rubber tube into the bottle and push it up to the top.
2. Now take a deep breath[6], place your lips round the free end of the glass tube (the mouthpiece) and breathe out as much air as you can.

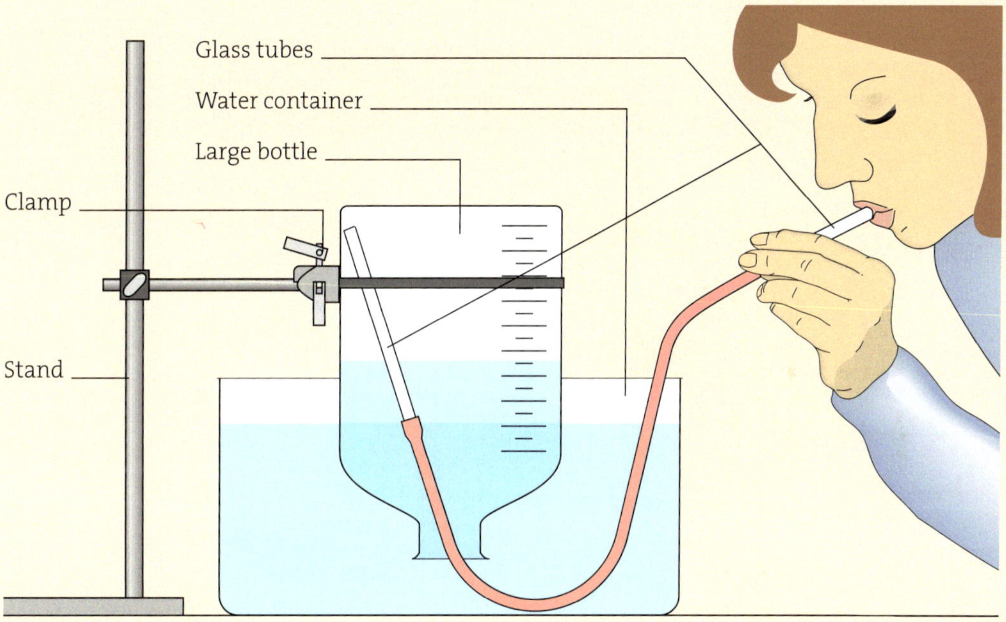

Figure 15 **Collecting the exhaled air**

3. Note what happens to the water in the bottle.
4. When you cannot exhale[7] any more, seal the open end of the bottle with your finger while pulling the tube out of the bottle. Now close the bottle with the stopper.
5. Record how much air the bottle contains. Keep the neck of the bottle under water the whole time.

1 measure sth. *etwas messen*
2 vital capacity *maximaler Lungeninhalt*
3 stopper *Stöpsel*
4 clamp *Klammer*
5 clamp sth. to sth. *etwas zusammenklammern*
6 breath *Atemzug*
7 exhale *ausatmen*

1 Describe what happened when you exhaled into the bottle.

2 Your teacher may ask different people in your class to carry out the experiment also. Among them there should be a good athlete and, if possible, also a pupil who plays a wind instrument (e.g. a clarinet, a saxophone or a trumpet), boys as well as girls. The mouthpiece should be exchanged or cleaned before each new person does the experiment.
Collect the results of everyone's vital capacities in a simple table and try to explain the differences.

3 What do you think you could do to increase the capacity of your lungs?

Activity 3 Finding out how exhaled air differs from inhaled[1] air

For this activity you need to know that lime water[2] is an indicator for carbon dioxide. The clear solution[3] becomes milky white when carbon dioxide dissolves[4] in it.

Materials:

- 2 Erlenmeyer flasks[5] (250 cm^3)
- 2 glass tubes, 15 cm long
- 2 glass tubes, 10 cm long
- a short glass tube for a mouthpiece
- 2 rubber bungs, each with two holes
- a T-tube
- 2 pieces of rubber tubing to connect the flasks with the T-tube
- a longer piece of rubber tubing to connect the mouthpiece with the T-tube
- freshly prepared lime water (about 300 cm)

Procedure:

1. Connect the two Erlenmeyer flasks with glass tubes and rubber tubing exactly as shown in Figure 16.
2. Fill each flask with 50 cm^3 of clear lime water.
3. Now place your lips round the mouthpiece and slowly breathe in and out 15 to 20 times. Do not breathe too hard!

1 inhale (sth.) *(etwas) einatmen*
2 lime water *Kalkwasser*
3 solution *Lösung*
4 dissolve *sich auflösen*
5 flask *Kolben*

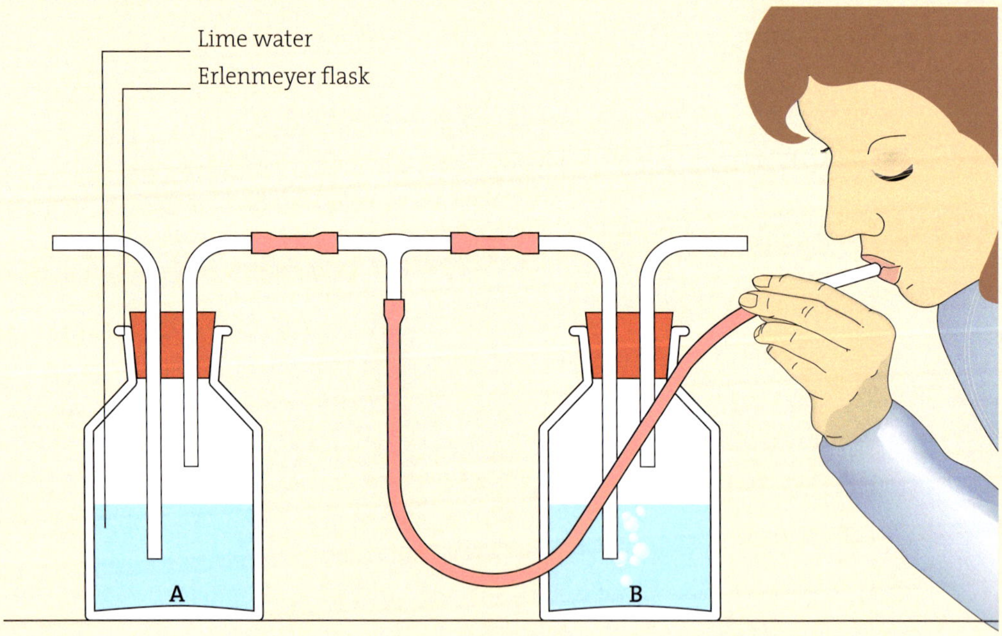

Figure 16 **Investigating the CO_2 content of inhaled and exhaled air**

1 In which flask did you see bubbles when you breathed out and in which did you see them when you breathed in?
2 Describe the special way in which the two flasks are connected.
3 In which flask did the lime water change?
4 What do the results of the experiment tell you about inhaled and exhaled air?
5 Explain the reasons for the difference between the air you breathe in and the air you breathe out.

Summary

Your respiratory system is made up of all the organs you use for breathing. When you draw air into your lungs, gases are exchanged in the alveoli: oxygen from the air passes into the blood while carbon dioxide from the organs passes from the blood into the lungs and is breathed out. The oxygen you take in travels to the organs, where it is used in a burning reaction to gain energy from food substances. In this process, carbon dioxide is formed and passes into the bloodstream.

Two different groups of muscles are used for breathing: the diaphragm and the intercostal muscles. When you are relaxed, you mostly use the diaphragm to expand your lungs. But when you need more oxygen, for example for physical exercise, the intercostal muscles help expand the chest so that the lungs can take in more air.

A Study tasks

1. Make a mind map based on the information in the summary above.
2. Write down all the words from the Key terms boxes on cards. With a partner, take turns drawing cards and explaining the terms. If you are not sure how to express something in English, check the text for words that can help you and copy them into your vocabulary notebook.
3. Draw a diagram showing the path of oxygen and carbon dioxide through your body. Copy it onto a transparency and present it to the class.

B Bio quiz

1. The following text has a lot of mistakes. Find and correct them.

> Air enters your body through your nose, where oxygen is taken up into the blood. It then flows through the windpipe, which is held open by rings of bone. In your lungs, the air goes through an artery until it arrives in the alveoli. Their walls are very thick, so oxygen passes out of the blood into the alveoli. The carbon dioxide is then transported to the organs in the red blood cells. Here it helps to get energy from sunlight, a process which is called external respiration.

2. Complete the sentences using **Key terms** from this module (see below).
   a) When we breathe in, the … inside our noses helps protect us against harmful substances.
   b) When you are sleeping, the movement of your … gives you enough air to breathe.
   c) The cells in our bodies use … to burn the food we eat, which produces … .
   d) Once you start smoking regularly it is hard to stop because the … in tobacco is very … .
   e) Laws against smoking in public places make it easier for non-smokers to avoid … .
   f) The most serious disease caused by smoking is …, which kills large numbers of people every year.

C Critical thinking

1 When you smoke, the cilia in your lungs stop moving for a while or at least slow down. What is the consequence of this?

2 When a rib is broken, there is a danger that the broken bone will tear the lung behind it. Explain what you think the result of this might be.

3 Research: The table on page 9 shows you non-smoking laws in some US states and in England. What is the current situation in Germany and how effective are the laws?

Key terms

See the boxes:

Anatomy of the lungs .... p. 4
Breathing (1) .... p. 4
Breathing (2) .... p. 5
Exchange of gases .... p. 6
Smoking and its effects .... p. 7
Breathing and speaking .... p. 13
Germs and infections .... p. 14

English-German vocabulary

A

air pollutant [pə'lu:tənt] *Luftschadstoff*

alveolus (pl. alveoli) [ˌælvi'əʊləs] *Lungenbläschen*

amputation [æmpju'teɪʃn] *Amputation*

B

bacterium (pl. bacteria) [bæk'tɪəriəm ☆ -'tɪriəm] *Bakterie, Bakterium*

be addictive [ə'dɪktɪv] *abhängig oder süchtig machen*

bell jar *glockenförmiges Gefäß*

blood vessel *Blutgefäß*

bloodstream *Blutbahn, Blutkreislauf*

branch *Ast, Zweig*

breath [breθ] *Atem, Atemzug*

breathing rate ['bri:ðɪŋ reɪt] *Zahl der Atemzüge pro Zeiteinheit*

breeding ground *Brutstätte*

bronchi ['brɒŋkaɪ ☆ 'brɑ:ŋ-] (sing. bronchus) *Bronchien*

bronchiole ['brɒŋkiəʊl ☆ 'brɑ:ŋ-] *Bronchiole*

bronchitis [brɒŋ'kaɪtɪs ☆ brɑ:ŋ-] *Bronchitis*

C

cancer ['kænsə] *Krebs, Krebserkrankung*

capillary [kə'pɪləri] *Kapillare*

carbon dioxide [daɪ'ɒksaɪd ☆ daɪ'ɑ:ksaɪd] *Kohlenstoffdioxid*

carbon monoxide [mə'nɒksaɪd ☆ mə'nɑ:ksaɪd] *Kohlenstoffmonoxid*

cartilage ['kɑ:təlɪdʒ] *Knorpel*

chemical compound ['kemɪkl 'kɒmpaʊnd ☆ 'kɑ:mpaʊnd] *chemische Verbindung*

cilium (pl. cilia) ['sɪliəm] *Wimperhärchen*

clamp *Klammer, Klemme*

clamp sth. to sth. *etwas zusammenklammern*

close-up *Nahaufnahme*

common cold *Schnupfen, Erkältung*

condition *Befinden, Zustand*

contract [kən'trækt] *sich zusammenziehen*

coordination [kəʊˌɔ:dɪ'neɪʃn ☆ koʊˌɔ:rd-] *Abstimmung, Koordination*

cotton wool *Watte*

cough [kɒf ☆ kɔ:f] *Husten*

D

diaphragm ['daɪəfræm] *Zwerchfell*

dissolve [dɪ'zɒlv ☆ -'zɑ:lv] *sich auflösen*

divide into sth. [dɪ'vaɪd] *sich in etwas aufteilen*

dome-shaped *gewölbt*

draw air *Luft ziehen*

drug *Droge, Arzneimittel*

E

emphysema [ˌemfɪ'si:mə] *Emphysem*

energy ['enədʒi] *Energie*

epiglottis [ˌepɪ'glɒtɪs ☆ -'glɑ:tɪs] *Kehldeckel*

Erlenmeyer flask *Erlenmeierkolben*

exhale *ausatmen*

external respiration [ɪk'stɜ:nl ˌrespə'reɪʃn] *äußere Atmung*

F

flask [flɑ:sk ☆ flæsk] *Glaskolben*

flu *Grippe*

foetus ['fi:təs] *Fötus*

fossil fuel ['fɒsl ☆ 'fɑ:sl 'fju:əl] *fossiler Brennstoff*

G

germ *Krankheitserreger*

get rid of sth. *etwas loswerden*

get used to sth. *sich an etwas gewöhnen*

glass tube [glɑ:s 'tju:b ☆ glæs 'tu:b] *Glasrohr*

gullet ['gʌlɪt] *Speiseröhre*

H

habit *Angewohnheit, Sucht*

hard palate ['pælət] *harter Gaumen*

headache ['hedeɪk] *Kopfschmerz*

heart attack *Herzanfall*

I

industrial plant *Industriebetrieb*

infect sb./sth. *jdn./etwas infizieren, anstecken*

infection *Infekt, ansteckende Krankheit*

inflamed *entzündet*

inflammation *Entzündung*

influenza [ˌɪnflu'enzə] *Grippe*

inhalation *Inhalieren*

inhale (sth.) *(etwas) einatmen*

inner lining *Schleimhaut*

intercostal muscles [ˌɪntə'kɒstl ☆ -'kɑ:stl 'mʌsls] *Zwischenrippenmuskeln*

internal respiration *innere Atmung*

investigate sth. *etwas untersuchen*

L

layer *Schicht*

left and right lungs *Lungenflügel*

lime water *Kalkwasser*

lung cancer ['lʌŋ kænsə] *Lungenkrebs*

M

measure sth. ['meʒə] *etwas messen*

moisten sth. ['mɔɪsn] *etwas anfeuchten*

mouth cavity *Mundhöhle*

mucous membrane ['mju:kəs 'membreɪn] *Schleimhaut*

mucus ['mju:kəs] *Schleim*

N

nasal cavity ['neɪzl 'kævəti] *Nasenhöhle*

nasal passage ['neɪzl 'pæsɪdʒ] *Nasengänge*

nicotine ['nɪkəti:n] *Nikotin*

nitrogen ['naɪtrədʒən] *Stickstoff*

O

oxygen ['ɒksɪdʒən ☆ 'ɑ:k-] *Sauerstoff*

P

passive smoking ['pæsɪv 'sməukɪŋ] *Passivrauchen*

pharynx ['færɪŋks] *Rachen*

physical exercise ['fɪzɪkl 'eksəsaɪz] *körperliche Aktivität(en)*

pitch *Tonhöhe*

pleural membrane ['plʊərəl 'membreɪn] *Lungen- und Rippenfell*

pneumonia [nju:'məʊniə ☆ nu:'moʊ-] *Lungenentzündung*

poisonous ['pɔɪzənəs] *giftig*

pregnancy ['pregnənsi] *Schwangerschaft*

R

record sth. [rɪ'kɔ:d] *etwas aufschreiben, notieren*

relax [rɪ'læks] *sich entspannen*

relieve [rɪ'li:v] *erleichtern*

respiration [ˌrespəˈreɪʃn] *Atmung*
respiratory system [reˈspɪrətri ☆ ˈrespərətɔːri] *Atmungssystem*
rubber bung *Gummistopfen*
rubber sheet *Gummihaut*
rubber tubing [ˈtjuːb- ☆ ˈtuːb-] *Gummischlauch*

S

settle *sich niederlassen*
soft palate [ˈpælət] *weicher Gaumen, Gaumensegel*
solution *Lösung*
sore throat [sɔː θrəʊt ☆ θroʊt] *Halsschmerzen, Halsentzündung*
spongy [ˈspʌndʒi] *schwammig, weich*
stopper *Stöpsel, Stopfen*
stroke *Schlaganfall*
suction bulb [ˈsʌkʃn bʌlb] *Saugball*
suction pump [ˈsʌkʃn pʌmp] *Saugpumpe*

T

tar *Teer*
tension [ˈtenʃn] *Spannung*
trachea [trəˈkiːə ☆ ˈtreɪkiə] *Luftröhre*
treat sb./sth. *jdn./etwas behandeln*
treatment *Behandlung*
tuberculosis [tjuːˌbɜːkjuˈləʊsɪs ☆ tuːˌbɜːrkjəˈloʊsɪs] *Tuberkulose*
tumour [ˈtjuːmə ☆ ˈtuːmə] *Geschwulst, Tumor*
twig *dünner Zweig*

V

vibrate [vaɪˈbreɪt ☆ ˈvaɪbreɪt] *vibrieren*
virus [ˈvaɪrəs] *Virus*
vital capacity [ˈvaɪtl kəˈpæsəti] *maximaler Lungeninhalt*
vocal cords [ˈvəʊkəl] *Stimmbänder*
voice box *Kehlkopf*

W

wet pack *feuchter Wickel*
wind-pipe *Luftröhre*

German-English vocabulary

A

abhängig machen be addictive
Abstimmung coordination
Amputation amputation
jdn./etwas anstecken infect sb./sth.
ansteckende Krankheit infection
etwas anfeuchten moisten sth.
Arzneimittel drug
Atem breath
Atemzug breath
Atmung respiration
Atmungssystem respiratory system
sich auflösen dissolve
etwas aufschreiben record sth.
ausatmen exhale
äußere Atmung external respiration

B

Bakterie/Bakterium bacterium (pl. bacteria)
Befinden condition
jdn./etwas behandeln treat sb./sth.
Behandlung treatment
Blutbahn bloodstream
Blutgefäß blood vessel
Blutkreislauf blood circulation
Bronchien bronchi
Bronchitis bronchitis

C

chemische Verbindung chemical compound

D

Droge drug

E

etwas einatmen inhale sth.
sich entspannen relax
entzündet inflamed
Entzündung inflammation
Erkältung (common) cold
erleichtern relieve
Erlenmeierkolben Erlenmeyer flask

F

feuchter Wickel wet pack
fossiler Brennstoff fossil fuel
Fötus foetus

G

Gaumen palate
gewölbt dome-shaped
giftig poisonous
Glasrohr glass tube
Grippe influenza, flu
Gummihaut rubber sheet
Gummischlauch rubber tubing
Gummistopfen rubber bung

H

Halsentzündung sore throat
Halsschmerzen sore throat
harter Gaumen hard palate
Herzanfall heart attack
Husten cough

I

Industriebetrieb industrial plant
jdn./etwas infizieren infect sb./sth.

Inhalieren inhalation
innere Atmung internal respiration

K

Kalkwasser lime water
Kapillare capillary
Kehldeckel epiglottis
Kehlkopf voice box
Knorpel cartilage
Kohlenstoffdioxid carbon dioxide
Kohlenstoffmonoxid carbon monoxide
Koordination coordination
Kopfschmerz headache
körperliche Aktivität(en) physical exercise
Krankheitserreger germ
Krebs(erkrankung) cancer

L

Lösung solution
etwas loswerden get rid of sth.
Luftröhre wind-pipe, trachea
Luftschadstoff air pollutant
Lungen- und Rippenfell pleural membrane
Lungenbläschen alveolus (pl. alveoli)
Lungenentzündung pneumonia
Lungenflügel left and right lungs
Lungenkrebs lung cancer

M

maximaler Lungeninhalt vital capacity
etwas messen measure sth.
Mundhöhle mouth cavity

N

Nahaufnahme close-up
Nasengänge nasal passage
Nasenhöhle nasal cavity
Nikotin nicotine
etwas notieren record sth.

P

Passivrauchen passive smoking

R

Rachen pharynx

S

Sauerstoff oxygen
Saugpumpe suction pump
Schicht layer
Schlaganfall stroke
Schleim mucus
Schleimhaut mucous membrane
Schnupfen (common) cold
schwammig spongy
Schwangerschaft pregnancy
Spannung tension
Speiseröhre gullet
Stickstoff nitrogen
Stimmbänder vocal cords
Stöpsel stopper
Sucht habit
süchtig machen be addictive

T

Teer tar
Tonhöhe pitch
Tuberkulose tuberculosis
Tumor tumour

U

etwas untersuchen investigate sth.

V

vibrieren vibrate
Virus virus

W

Watte cotton wool
weicher Gaumen soft palate
Wimperhärchen cilium (pl. cilia)

Z

sich zusammenziehen contract
Zustand condition
Zweig small branch, twig
Zwerchfell diaphragm
Zwischenrippenmuskeln intercostal muscles

Abbreviations and symbols

cm	centimetre(s)
cm^3	cubic centimetre(s)
e.g.	(Latin) *exempli gratia* = for example
g	gram(s)
jdn.	*jemanden*
p., pp.	page, pages
pl.	plural
sb.	somebody
sth.	something
☆	American pronunciation follows

Regular differences in American pronunciation – as for instance the [r] after vowels – have not been given.

Acknowledgements

Bildquellenverzeichnis
Alamy images, Abingdon: S. 7 (l. o. © Paul Rapson, l. u. © Ashley Cooper), S. 8 o. (© Medical-on-Line); V. Döring, Hohen Neuendorf: S. 7 r. o., S. 11 u.; Gabriele Heinisch, Berlin: S. 5; Institut für Plastination, Heidelberg: S. 8 u. (beide); medicalpicture GmbH, Köln: S. 13 u.; Lutz-Erich Müller, Leipzig: S. 4, S. 6, S. 10, S. 11 o., S. 13 o., S. 15, S. 16; Photolibrary, London: S. 14 (© Matthias Stolt): Shutterstock images, New York (© Sebastian Kaulitzki).